The Tower of Pisa

T0172585

Built Heritage and Geotechnics

Series editor: Renato Lancellotta

Volume III

ISSN: 2640–026X
eISSN: 2640–0278

The Tower of Pisa

History, Construction and Geotechnical Stabilization

J.B. Burland
Imperial College, London

M.B. Jamiolkowski
Politecnico di Torino, Italy

N. Squeglia
Università di Pisa, Italy

C. Viggiani
Università di Napoli, Italy

CRC Press
Taylor & Francis Group
Boca Raton   London   New York   Leiden

CRC Press is an imprint of the
Taylor & Francis Group, an **informa** business

A BALKEMA BOOK

CRC Press/Balkema is an imprint of the Taylor & Francis Group, an informa business

© 2021 Taylor & Francis Group, London, UK

Typeset by Apex CoVantage, LLC

Library of Congress Cataloging-in-Publication Data
Names: Burland, J. B., author. | Jamiolkowski, M. B., author. | Squeglia, Nunziante, author. | Viggiani, Carlo, author.
Title: Tower of Pisa : history, construction and geotechnical stabilisation / J.B. Burland, Imperial College, London, M.B. Jamiolkowski, Politecnico di Torino, N. Squeglia, Università di Pisa, C. Viggiani, Università di Napoli.
Description: First edition. | Boca Raton, FL : CRC Press/Taylor & Francis Group, [2020] | Series: Built heritage and geotechnics, 2640-026X ; 3 | Includes bibliographical references.
Identifiers: LCCN 2020011471 (print) | LCCN 2020011472 (ebook)
Subjects: LCSH: Leaning Tower (Pisa, Italy)—History. | Pisa (Italy)—Buildings, structures, etc. | Architecture—Italy—Pisa—Conservation and restoration.
Classification: LCC NA5621.P716 B87 2020 (print) | LCC NA5621.P716 (ebook) | DDC 720.945/551—dc23
LC record available at https://lccn.loc.gov/2020011471
LC ebook record available at https://lccn.loc.gov/2020011472

Published by: CRC Press/Balkema
          Schipholweg 107C, 2316 XC Leiden, The Netherlands
          e-mail: Pub.NL@taylorandfrancis.com
          www.crcpress.com – www.taylorandfrancis.com

ISBN: 978-0-367-46904-7 (hbk)
ISBN: 978-1-003-04612-7 (ebk)

DOI: 10.1201/9781003046127
DOI: https://doi.org/10.1201/9781003046127

Contents

Foreword

This is the third of a series of volumes on Built Heritage and Geotechnics, intended to reach a wide audience: professionals and academics in the fields of civil engineering, architecture and cultural resources management, particularly those involved in art history, history of architecture, geotechnical engineering, structural engineering, archaeology, restoration and cultural heritage management, and even the wider public.

Motivations of this series rely on the fact that preservation of built heritage is one of the most challenging problems facing modern civilization. It involves inextricable patterns in various cultural, humanistic, social, technical and economic aspects. The complexity of the topic is such that a shared framework of reference is still lacking among art historians, architects, and structural and geotechnical engineers. It is sadly the case that, despite some superb examples of the appropriate and respectful conservation of valuable ancient monuments, of which the Pisa Tower is one, nevertheless cases persist of the replacement of treasured historic monuments with new ones, which are nothing but pale modern imitations of the original monument.

For these reasons, publishing short books on specialized topics – like well documented case studies of restoration works at specific sites or monuments or of detailed overviews of construction techniques, intended as material witnesses of knowledge of the historical periods in which the monuments were built, or specific conservation works – may be of great value.

The present volume is about the Leaning Tower of Pisa, no doubt the favorite "shorthand" image for the idea not only of Pisa but of Italy. However, the Tower is just a single component of Pisa's amazing religious core, the so-called Campo dei Miracoli, that should be better addressed as Campo delle Mirabilia, being a sight whose impact no amount of prior knowledge can blunt.

The buildings date from the period of Pisa's greatest prosperity and power, from the 11th to the 13th centuries: the Cathedral was begun in 1063 and completed at the end of the 12th century; the Baptistery was started in 1152; the bell tower in 1173; and the cemetery, the Camposanto, was added at the end of the 13th century. The Pisan Romanesque architecture of this period, distinguished by its white-and-black marble façades, is complemented by the impressive and finest medieval sculpture from the workshops of Nicola and Giovanni Pisano. And the impressive artistic and religious unity of this complex is such that the Tower would have been astonishing even if not leaning.

Within this context, this book is a fascinating narration of the history of the construction of the Leaning Tower, the techniques involved and the attempts that masons made during construction to mitigate the tilt from the very beginning. Finally, it describes a rather unique solution that is today known as *underexcavation* in the technical literature: soil was removed at selected locations below the foundations in order to induce settlement and to stop the tilt that was increasing at an alarming rate.

The idea of underexcavation for the Pisa Tower was first suggested by Leonardo Terracina, as published in *Géotechnique* in 1962; however, appropriate historical precedents were the 19th-century interventions on leaning bell towers performed in England and in Holland. More recently the method had been successfully applied to correct the differential settlements of the Metropolitan Cathedral of Mexico City, as described in a companion book of this series.

The stabilizing measures applied to the Tower "without even touching it" reveal, in addition to their technical and scientific aspects, an important cultural issue that must be recalled: quite often the integrity requirement is interpreted only as the requirement of preserving the shape and the appearance of the monument. In reality, it also implies historic integrity, that is, consideration of the changes the monument experiences with time, as well as its material integrity, meaning construction techniques, materials and structural scheme. Therefore, preserving integrity requires a multidisciplinary approach, as well as the attitude not to rush into deciding on stabilizing measures until the behavior and history of the monument are properly understood. The case history of the Pisa Tower is a powerful exemplar because it reveals that, only through a deep and patient scrutiny of the measurements of the movements of the Tower made since 1911, was it possible to identify an unexpected mechanism of foundation movement consistent with the phenomenon of "leaning instability."

Leaning instability occurs when a tall structure reaches a critical height at which the overturning moment generated by a small increase in inclination is equal to or larger than the resisting moment generated by the foundation. Leaning instability is due not to a lack of strength of the ground but to insufficient stiffness, and the recognition of this phenomenon proved to be crucial in developing stabilization measures.

In short, as the art historian Salvatore Settis has observed, "[A] variation in inclination was already dictated by the Tower's genetic code."

Is there anything else to be added? Let's leave the reader to absorb in depth and enjoy the contents of this book, written in an accessible way by the experts in charge of the stabilizing project.

Renato Lancellotta, Series Editor

Chapter 1

Introduction

The town of Pisa, located on the Arno river near the Tyrrhenian coast of Italy, became a flourishing commercial center and a powerful maritime republic in the 11th century. Having defeated the Saracens in a series of sea battles, it grew into a very important commercial and naval center, took control of the Mediterranean sea and acquired colonies in Sardinia, Corsica, Elba, Southern Spain and North Africa (Fig. 1.1). In the 12th century, it was the naval base for the First Crusade in which a fleet of 120 Pisan ships participated; it established a number of settlements in the Holy Land, founding colonies in Antioch, Acre, Jaffa, Tripoli, Tyre and Larakia.

It was in the period of maximum splendor of the Republic, in the 12th and 13th centuries, that the monuments in Piazza dei Miracoli (Miracles Square) were erected. The Square, with the Cathedral, the Baptistery and the Leaning Tower (Fig. 1.2), is the awesome manifestation of the ideal unity that reigned at the time among religious, spiritual and political powers. In its monuments, civil history and history of art intertwine, giving them an extraordinary character as sign and symbol of the city (Franchi Vicerè, Viggiani et al., 2005; Franchi Vicerè, Veniale et al., 2005). Civic pride, identity and a sense of belonging are evident in an engraving on a stone on the façade of the Cathedral, recalling in epic tone that the treasures captured from the Saracens, after taking Palermo harbor in 1063, initially funded the construction.

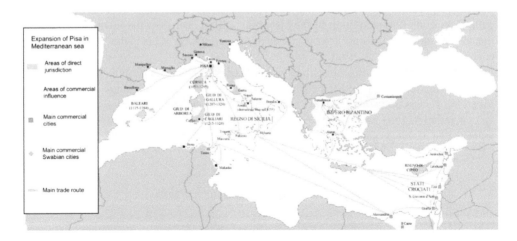

Figure 1.1 Commercial and territorial expansion of the Republic of Pisa in the 12th century: trade routes, colonies and warehouses

Figure 1.2 Pisa, Piazza dei Miracoli: the Leaning Tower, the Cathedral and the Baptistery

About one century later, in 1172, a lady named Berta, widow of Calvo, left to the Opera della Primaziale (the institution of the bishopric, still in charge of the monuments) a testamentary bequest of 60 coins "for the stones of the bell tower." This is the first known reference to the Tower.

More than eight centuries later, the authors became the geotechnical team of an International Committee appointed by the Italian Government "for the Safeguard of the Tower of Pisa." In 11 years of difficult, intense and stimulating work, the Committee succeeded in the task of stabilizing the monument while at the same time respecting its integrity. This book tells the tale of this project.

Chapter 2

The Leaning Tower

The Tower is composed of a cylindrical masonry body surrounded by loggias with arches and columns that rest on the base cylinder and that are surmounted by a belfry. The structure is subdivided into eight segments, called "orders" (Locatelli *et al.*, 1971). A vertical section on the plane of maximum inclination (the direction of maximum inclination is very close to being north–south) is shown in Figure 2.1. The monument is 58.4 m in height from the

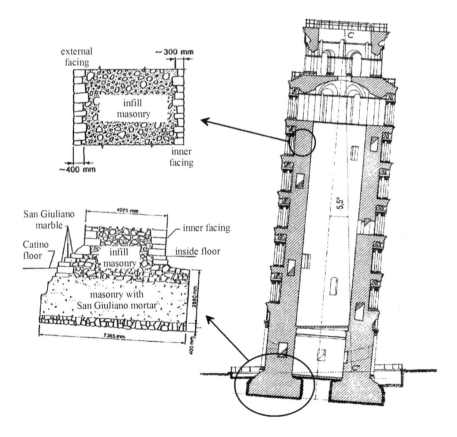

Figure 2.1 Cross section of the Tower with the plane of maximum inclination, the direction of the inclination being very nearly southward

plane of the foundation, rising above ground level for more than 55 m. Its weight has been estimated at 142 MN; the center of gravity is 22.6 m above the foundation plane. The ring foundation has an external diameter of 19.6 m; the center hole, 4.6 m. The foundation area is thus 285 m², the average pressure exerted on the ground is 497 kPa. In 1993, the inclination was 5.5° (in other words, about 10%); the corresponding value for eccentricity in the foundation was 2.3 m.

The central body is a hollow cylinder made up of two facings in dressed stone and, between them, a zone filled with rubble masonry of lime mortar and irregularly shaped stones. Inside this zone, there is the spiral stairwell and stairway, which leads to the belfry with its 293 steps.

Evaluations of the state of stress in the masonry (Macchi and Ghelfi, 2005) show that the most highly stressed section is that on the south side just above the level of the first cornice, where the cross section of the masonry undergoes an abrupt reduction. What is more, the section is weakened by the presence of the staircase and door openings; in this zone, the stress in the external facing on the south side has been evaluated at 8 MPa (Fig. 2.2). The fissure pattern is not easy to interpret due to the repairs that have been carried out over the whole life span of the monument; nevertheless, more fissures and cracks are apparent on the south side of the structure than on the north. This situation aroused many worries concerning the structural safety of the Tower and led, in 1990, to the decision to close it to visitors. What

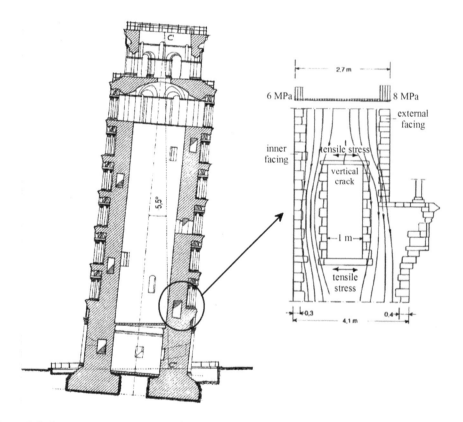

Figure 2.2 Stress concentration in the masonry

was feared was a brittle failure of the masonry and phenomenon of local instability in the most highly stressed areas of the external facing on the south side and at the level of the first cornice. Due to the fragility of the materials, this type of phenomenon could have caused the entire Tower to collapse almost without warning.

The section in Figure 2.1 shows the excavation that today runs around the base of the Tower (the so-called *catino*). It was added in 1838 to expose the lowest portion of the Tower, until that time hidden in the ground due to a settlement of at least 3 m.

Finally, the Tower is not straight, but its axis is curved, with the concavity toward the north. This characteristic may be perceived in the exaggerated drawing of Figure 2.3; it is described, quite irreverently, as the "banana shape" of the Tower.

Figure 2.3 So-called banana shape of the Tower

Chapter 3

The subsoil of the Tower

Figure 3.1 shows a schematic north–south section of the ground underlying the Tower (Viggiani and Pepe, 2005). The subsoil of Piazza dei Miracoli, like the entire plain of Pisa, consists of geologically recent lagoon and marsh deposits (Pleistocene–Holocene). These are silts, clays and fine sands intercalated with the wind-borne sands of the ancient coastal dunes.

From the ground surface (that is located at $2.5 \div 3$ m above the mean sea level) downward, we encounter three main layers with different geotechnical properties.

Layer A is about 10 m thick. After $2 \div 3$ m of ground made with various archaeological remains, it consists mainly of estuarine deposits, laid down under tidal conditions. The soils are therefore rather irregularly layered sands, silts and clays. At the lower limit of this complex, in contact with the underlying clays, we find a 2-m-thick layer of fine, medium dense sand, gray in color. Furthermore, detailed examination of boreholes and penetrometer profiles reveals that south of the Tower, the finer fractions are prevalent, and the sandy layer tends to thin. Since soils with finer grains are also slightly more compressible, this factor probably initiated the southward inclination of the Tower. Fine-grained soils, however, deform slowly because of the consolidation process; this could explain why, initially, the lean of the Tower was northward.

Layer B extends to a depth of about 40 m below ground level and is constituted mainly of marine clays. The *layer* is subdivided into four distinct sub-layers. The upper clays, locally known as the *pancone*, are slightly overconsolidated to normally consolidated soft to medium sensitive clays; this layer is the principal cause of the foundation problems of the Tower. Below this, an intermediate layer of stiffer clay is found, which, in turn, rests upon a sand layer, the intermediate sands. The bottom layer of *Layer B* consists in normally consolidated, medium to stiff lower clay.

Layer C is a dense sand, reaching to a depth of at least 70 m.

Around the Tower, the surface of separation between the upper sands and the clays of the *pancone* lies on a horizontal plane with maximum deviations of just a few centimeters, except for the depression right underneath the Tower, with a depth of about 2.2 m. This depression is evidently a deformation induced by the weight of the Tower and is one of the clues to reconstructing the average settlement of the Tower. It is, in fact, equal to 2.2 m plus the compression of the soil between the foundations of the Tower and the top of the *pancone*. From this and other studies, as previously mentioned, the settlement of the Tower has been evaluated to be not less than 3 m.

The groundwater (Fig. 3.2) in *Layer A* is at a depth of between 1 and 2 m, that is, at an average piezometric level close to 1.5 m a.s.l.

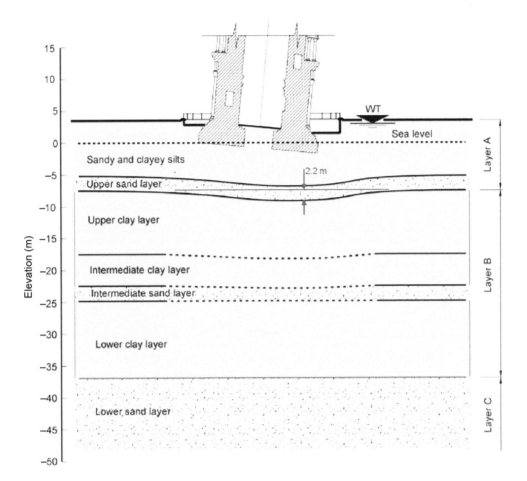

Figure 3.1 Subsoil of the Tower

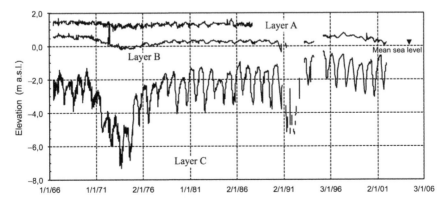

Figure 3.2 Groundwater regime in the subsoil of the Tower

In the deep sands, the piezometric level is on the average about 2 m below sea level; that is, about 3.5 m lower than that of the groundwater in *Layer A*. It undergoes cyclic fluctuations, with an annual period and amplitude of about 4 m (±2 m with respect to average value). This trend is linked to the withdrawal of water, for irrigation and industrial use, from the subsoil of the whole Pisa plain by means of wells. The result, in *Layer B*, is a downward seepage and piezometric levels slightly lower than hydrostatic.

In the early 1970s, due to more intense pumping connected to a sequence of dry years, the piezometric level in the lower sands decreased to more than 6 m below the mean sea level; as we will see, this drawdown produced an increase of the rate of rotation of the Tower. The closure of several wells in the Square and its vicinity arrested this trend.

Chapter 4

History of the construction and inclination

Is the inclination of the Tower an intentional feature of the monument, a sort of virtuosity of the ancient (and unknown) architect, or the results of an accident? This question was the subject of a long and heated debate in Pisa in the 19th century, with the large majority inclined to support the former hypothesis. Another school of thinking, headed by the architect Alessandro Gherardesca (we will meet this man again later), was of the opposing opinion. The answer to this question is written in the Tower itself and in its history, which we briefly recall here (Lumini and Polvani, 1971).

Work on the Tower began in 1173 (Fig. 4.1); as previously mentioned, the architect is not known with certainty, but the current attribution is to Bonanno Pisano, together with a Wilhelm from Innsbruck. This attribution has been repeatedly criticized and was considered outdated, but it has been recently strongly supported by a fine essay of Ammannati (2018).

Construction had progressed to the 4th order by 1178, when work was interrupted. The reason for the stoppage is not known but, had the construction continued much further, the Tower would have experienced a collapse (Burland and Potts, 1994; WESI Geotecnica, 2016; Leoni *et al.*, 2018). Modern geotechnical engineers have been fascinated by the idea that the ancient masons had an understanding of soil mechanics and implemented a stage construction to take advantage of the pore pressure dissipation and increased rigidity and strength of the soil. On the contrary, no trace of even a minimum worry about the statics of the Tower can be found in the very rich documentation available in the archives about the construction of the monument. The reason for the interruption is probably to be found in a shortage of resources: at the time, Pisa was at war with Genoa, Lucca and Florence and engaged in other constructions such as the Baptistry, the Camposanto and the Hospital. One cannot but conclude that fate was at play!

After a pause of nearly 100 years, in 1272, the architect Giovanni di Simone resumed the work. By 1278, construction had reached the 7th cornice, when it was interrupted again for 90 years. As before, had the work continued, the Tower would have fallen over. Again, this was not the reason for the interruption, even if the masons were aware of the inclination (they were correcting for it, producing the banana shape!). A few years later, in 1298, a commission inspected the Tower, measured the inclination and left a formal record of the operation.

On Wednesday March 15 the wise men Master John, son of Nicholas, Mason, Master Guy, son of John, Mason and Master Ursellus, Woodworker, together plumbing by common consent the bell tower of the Pisan cathedral by means of a plumb line, from the top to the bottom, agreed in the presence of me, the notary that plumb, hanging from the wire, touched the ground in a place that they marked unanimously.

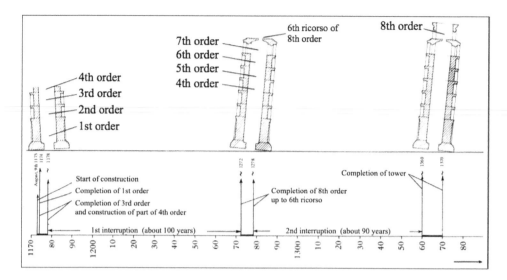

Figure 4.1 Schematic history of the construction of the Tower

> Recorded in Pisa in that place, in the presence of sir Guelfo, canon of Pisa, of Nerio cleric son of Guy, and Ceccho clergyman of Pisan Chapter and many other witnesses.
> (Archives of Pisan Chapter, Proceedings, 1298–1306)

Again, there are no traces of concern about the inclination. In this connection, however, it must be pointed out that, in the Middle Ages, leaning towers were a usual feature of the urban skyline in Italy.

The reasons for the interruption have to be found again in political and military problems. Pisa, as a powerful state, was destroyed forever by the crushing defeat of its navy in the battle of Meloria against Genoa in 1284. In this battle, most of the Pisan galleys were destroyed and its mariners taken prisoners. Only a small group of galleys, under the command of the famous count Ugolino della Gherardesca, escaped. For this, Ugolino was accused of treason and enclosed in a tower with his sons; the keys were thrown into the Arno river and the prisoners left to starve.

The bell chamber was commenced around 1360 by Tommaso di Andrea Pisano and completed around 1370, two centuries after the start of the construction.

We know that the Tower must have been tilting right from the beginning of construction because the ancient masons corrected for the inclination, eventually giving the Tower its banana shape. The most important correction had to be applied to the belfry, due to the inclination developed during the second interruption of the work. On the north side, indeed, there are only four steps from the seventh cornice up to the floor of the bell chamber, while on the south side there are six steps and five on the east and west sides (Fig. 4.2).

The adjustments made during construction, by inserting tapered masonry layers of gradually changing thickness, are a reliable clue to the history of the tilt. Based on the shape of the axis of the Tower and on a hypothesis on the way the masons corrected for the progressive

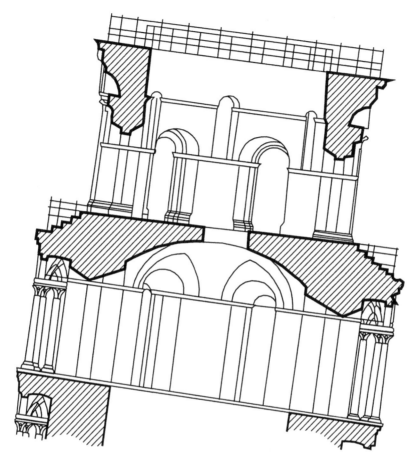

Figure 4.2 Last correction: at the top of the 7th order, there are six steps on the south and only four on the north

lean of the Tower, the history of the inclination of the Tower foundation reported in Figure 4.3 may be deduced (Burland and Viggiani, 1994).

During the first phase of construction, to just above the 3rd cornice (1173–1178), the Tower inclined slightly to the north. The construction stopped then for a century, and when it recommenced in 1272, the Tower began to move southward. When the construction reached the 7th cornice in 1278, the inclination was about 0.6° southward. During the next 90 years, the construction was again interrupted, and the inclination increased to about 1.6°, at an average rate of 40 arc seconds per year[1] (in terms of overhang, about 10 mm per year, a rate much faster than that at the end of the 20th century). After the completion of the bell chamber in about 1370, the inclination went on increasing.

Pictures and documents may furnish some information, among them a fresco painted in 1385 by Antonio Veneziano (Fig. 4.4) and the value reported by Giorgio Vasari in his *Life of Arnolfo di Lapo* (Vasari, 1550).

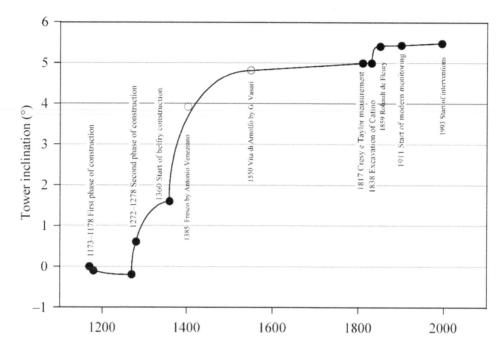

Figure 4.3 History of the inclination of the Tower: 1173–1370, values of inclination deduced from the shape of the axis; 1370–1817, from documents and pictures; since 1817, from direct measurements

In 1817, when two English architects, Cresy and Taylor (1829), made the first recorded measurement, the inclination of the Tower was about 4.9°. The average rate of increasing inclination from 1550 (Vasari's value) to 1817 equals 2.7 arc seconds per year.

Admittedly, this is a very rough estimate, due to the uncertainty of the Vasari's value, but it suggests that, in the early 19th century, the Tower was nearly motionless.

A French architect, Rohault de Fleury (1859), carried out another measurement 40 years later, finding a value of the inclination significantly higher than Cresy and Taylor's.

Between the two measurements, another important event in the history of the Tower had occurred. The architect Alessandro Gherardesca excavated the walkway around the base of the Tower, known as the *catino* (the basin), with the purpose of exposing the column plinths and foundation steps for all to see as originally intended. As previously mentioned, the base of the Tower had sunk into the soil due to the 3-m settlement.

The excavation of the *catino* produced a sudden increase in the inclination of the Tower but also a variation of the characteristics of its motion. Before the excavation, the Tower had come to rest, or, in any case, its motion was going on at a very low and progressively decreasing rate. After the excavation, the Tower moved at a progressively increasing rate, ineluctably destined to end in collapse. The excavation by Gherardesca is an example (not the only one, as we will see) of the danger of good intentions. They are no guarantee of success; as it is said, the road to hell is paved with good intentions. Incidentally, today even an elementary knowledge of soil mechanics explains that an excavation around a shallow

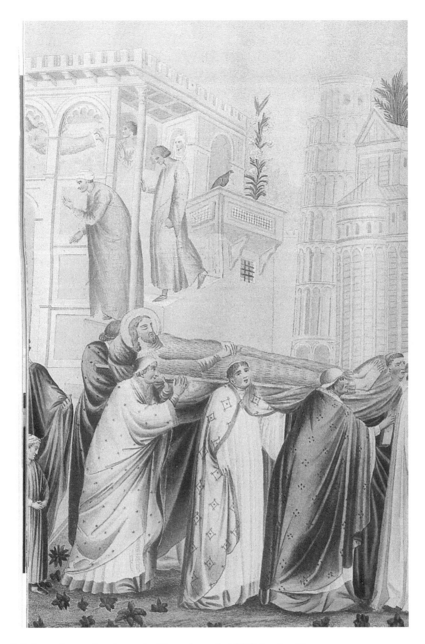

Figure 4.4 Pisa: fresco by Antonio Veneziano (~1385) in the Camposanto

foundation significantly decreases its safety. Gherardesca was moved by strong enthusiasm and by the will to bring about a better fruition of the monument, but he came very close to bringing about the collapse of the Tower.

Since 1911, the inclination of the Tower has been monitored by different means; the results of these observations will be reported in the next chapter.

Note

1 For the Tower, a change of inclination of 1 arc second is equivalent to a horizontal displacement at the top equal to approximately 0.3 mm.

Chapter 5

The results of monitoring, 1911–1990

The movements of the Tower during the 20th century have been observed by means of a comprehensive monitoring system, which was started in the early 1900s and has been progressively improved (Burland and Viggiani, 1994; Viggiani and Squeglia, 2005a, 2005b; Viggiani *et al.*, 2005a). Figure 5.1 reports the instruments of geotechnical interest; in addition, a number of instruments to observe the behavior of the masonry have been installed (deformometers on the fissures, thermometers and sensors of direction and velocity of the wind, air temperature, sun irradiation, accelerometers).

Geodetical measurements have been employed at least once per year since 1911. The GB pendulum was installed in 1934, and since then it has been observed daily, with an interruption in 1942–1948 due to World War II. Four leveling points on the base of the Tower were

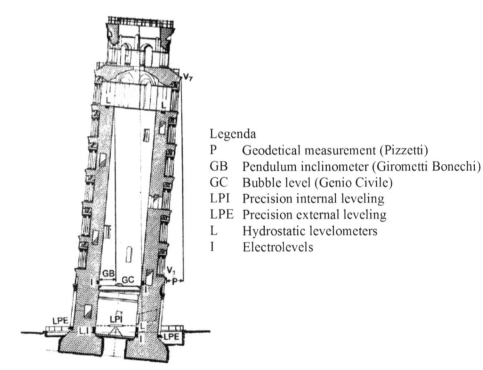

Legenda
P   Geodetical measurement (Pizzetti)
GB   Pendulum inclinometer (Girometti Bonechi)
GC   Bubble level (Genio Civile)
LPI Precision internal leveling
LPE Precision external leveling
L   Hydrostatic levelometers
I   Electrolevels

Figure 5.1 Monitoring system of the Tower

installed in 1928; in 1965, 15 external points. In 1993, eight further internal points were added; they are used currently for the precise determination of the rotation of the Tower foundation.

The increases in the inclination of the Tower since 1911 are reported in Figure 5.2. From 1911 to 1934, only geodetical measurements are available; after 1934, different types of measurements have been added.

Figure 5.2 shows that the inclination of the Tower has been steadily increasing. There is a clear trend of the inclination of the whole Tower (as measured by the GB pendulum and by geodetic survey) to increase more than the inclination of the foundation (as measured by precision leveling near the base of the Tower and by GC bubble level). This is the effect of a deformation of the Tower body.

The long-term trend of rotation is affected by some disturbances: grouting of the foundation body and of the soil surrounding the *catino* in 1935; boreholes in the masonry and the subsoil in 1966 and 1985; intense water withdrawal from the lower sands in the early 1970s. In any case, the rate of rotation of the Tower was increasing: about 4 arc seconds per year in the 1930s, about 6 arc seconds per year in the 1990s; this latter figure is equivalent to a horizontal movement at the top of about 1.5 mm per year (Jamiolkowski, 2005). There has been much debate about the cause of this progressive increase of the inclination. Usually, it has been attributed to creep in the underlying soft clay of the *pancone*, the assumption being made that the south side was settling more than the north side; we will come back to this point.

Figure 5.3 reports in detail the increase of the inclination between 1986 and 1991, as measured by the GB pendulum at 9 a.m. each day. Cycles of increase and decrease of the inclination superimpose to the long-term trend of 6 arc seconds per year; a similar behavior is observed in the orthogonal east–west plane but without a long-term increase. Some interesting elaborations are reported in Figure 5.4.

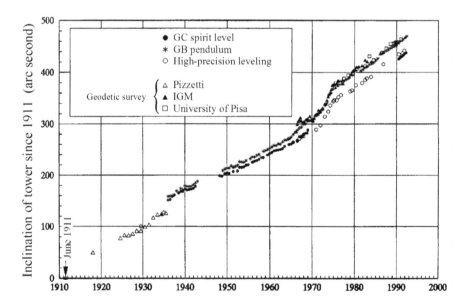

Figure 5.2 Increase of the inclination of the Tower since 1911

Figure 5.3 Increase of the inclination of the Tower in the north–south plane, as measured daily at 9 a.m. by the GB pendulum inclinometer

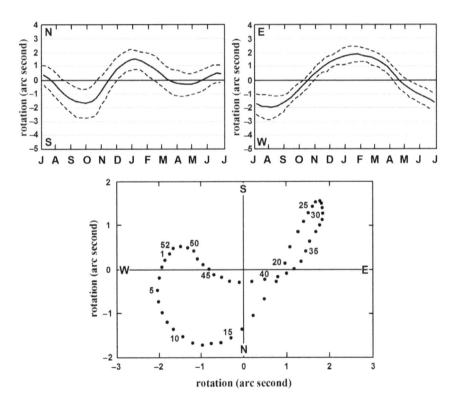

Figure 5.4 Cyclical variations of the inclination of the Tower in a year. In the north–south direction, the diagram has been obtained after having subtracted the long-term permanent increase. Full line represents the average of the observations in the period 1935–1992; data are smoothed with a mobile averaging over 15 days. Dotted lines represent the boundaries of a band with a width equal to twice the standard deviation (Probabilitas, 1992). The third diagram is derived from the other two using weekly intervals of time.

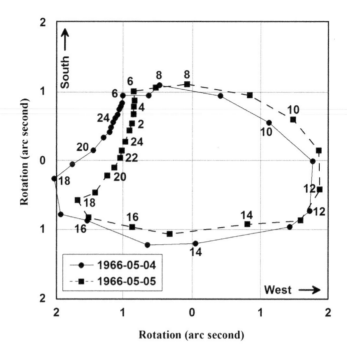

Figure 5.5 Variation of the inclination of the Tower over two days (May 4/5, 1966), as determined by GB pendulum inclinometer

Going further in detail, in Figure 5.5, the variation of the inclination in the span of two days is reported. The inclination has a daily cycle, almost entirely reversible. This movement is due to the heating of the external part of the Tower when exposed to the sun (Noccioli *et al.*, 1971).

Chapter 6

Studies and undertakings in the 20th century

The first Commission on the Tower of Pisa installed by the Italian Government was a consequence of the worries induced among the public by the 1902 collapse of the S. Marco bell tower in Venice. The Commission carried out a number of investigations and presented the results in a broad and valuable report, issued in 1913 (Ministero P.I, Opera Primaziale, 1913).

A second Commission was installed in the same year with the task of studying the possible means of stabilizing the Tower but did not conclude because of World War I. A new Commission with the same task was nominated in 1924; it included a number of experienced engineers and developed a solution consisting in widening the foundation of the Tower by filling the *catino* with concrete. The proposal met with strong opposition in Pisa, and local authorities appointed an alternative Commission to develop different and less intrusive solutions.

In 1927, the Government succeeded in unifying the two Commissions into a new one, which came to the conclusion that the most urgent need was that of sealing the *catino*. As mentioned, since 1838 the walkway was kept dry by continuous pumping, and it was believed that the resulting seepage, with possible fine erosion, could be the cause of the continuous increase of the inclination. In the years 1934–1935, the Tower foundation and the soil surrounding the *catino* were made watertight by injecting 100 t of cement grout into the foundation masonry and 21 m^3 of chemical grout into the soil. As already reported, the intervention stopped the water inflow into the *catino*; the price for this success, however, was a new, sudden and marked increase of the inclination of the Tower. About 100 years after the excavation of the *catino*, another intervention carried out with wishful thinking to stabilize the Tower had again strongly threatened it; this confirms that the road to hell is paved with good intentions!

In July 1944, during World War II, Pisa was divided: Germans were resisting to the north, while Americans advanced from the south. Every bend of road, every farmhouse and every escarpment seemed to be occupied by obstinate German defenders. As the number of deceased and wounded Americans mounted, the advance was in danger of stalling. How could the German artillery be so accurate in such a flat, coastal terrain? They had to have a vantage point – maybe it was the Leaning Tower? Sergeant Leon Weckstein, of the US 91st Infantry Division, was entrusted the most important mission of his war: to get close enough to the Tower to find out if the Germans were inside. If any enemy activity had been detected, Americans were not going to sacrifice men for a chunk of masonry, no matter how old.

Many years later, Weckstein (1999) published a book of war memories. "I took my time," he writes in his book, "training the binocular slowly up and down, attempting to discern anything that might be hidden within those recesses and arches." But after a whole day of observation, he did not call down fire. Waiting for the signal were inland gun batteries and a destroyer offshore.

What the 91st Infantry Division did not know was that they were entrusting one of the war's most fateful missions to a man rejected by the Navy for being shortsighted. "In 1942 the Navy told me to go away and eat carrots for six months," writes Weckstein. "Then the Infantry took me – but they take anyone!" It is not known whether the Tower was in fact occupied by the Germans as an observation point – Fate at play again!

After World War II, it became clear that the Tower was still moving, in spite of the work carried out in 1935. A permanent Commission was thus appointed in 1949, and, among other tasks, it had to examine and evaluate a number of design schemes. Though proposed by renowned engineers, all of the approaches were intrusive and disrespectful of the historical and material integrity of the monument; with hindsight, it appears very lucky that the Commission did not recommend any of these solutions. Though being "permanent," the Commission was dismantled in 1957; only the inclination measurements were continued. Further solutions were proposed in the following years; it is noteworthy that one of the most intrusive (Fig. 6.1) was suggested by the architect N. Benporad, Superintendent of Monuments of Pisa! Fortunately, none of these proposals was taken into consideration, and the integrity of the Tower was saved. Fate was always at play.

In 1964, a new and very important Commission was appointed, with the task of preparing the documents of an international competition for the design and implementation of stabilizing works. The Polvani Commission, named after its chairman, included for the first time a group of geotechnical engineers: C. Cestelli Guidi, University of Rome; A. Croce, University of Napoli; E. Schultze, University of Aachen; and A.W. Skempton, Imperial College London. To make complete documentation available, the Polvani Commission carried out a number of investigations and collected an impressive amount of knowledge (Ministero LL.PP., 1971).

The call for tender was issued in 1973. Twenty-two groups participated in the competition, and 11 among them were admitted. Professor Polvani had died in 1970, but the Commission, with minor modifications, was charged with judging the competition. Five proposals were judged worthy of mention (Fig. 6.2), but eventually no contract was awarded. At that time, it was discovered that the Piazza dei Miracoli was experiencing subsidence induced by pumping water from deep wells (Croce *et al.*, 1981). This factor was not properly considered in the tender, and this was one of the reasons why the contract was not awarded. Three of the groups mentioned joined together in a Consorzio and developed a common solution working in connection with the Commission, but eventually nothing was done.

In 1983, the Ministry of Public Works installed a Design Group with the task of designing the stabilization work; they produced a very sophisticated but still rather intrusive solution that was not definitively approved by the Council of Public Works. In 1988, a technical Committee, entrusted by the Government to study the problem, focused attention on the risk of a brittle failure of the heavily stressed masonry, in addition to the risk of foundation failure. A failure of the masonry would be sudden, without forewarnings, and therefore potentially very dangerous.

A confirmation of this danger was another spectacular collapse of a historic tower, which occurred in Italy in 1989: that of the Civic Tower of Pavia, with four fatalities. As a result, attention to the safety of the Tower of Pisa increased, and the Government prohibited access to visitors, following a recommendation of the technical committee.

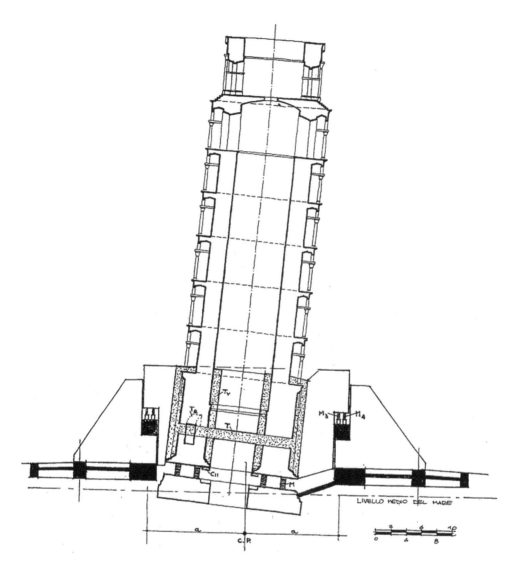

Figure 6.1 Proposal by Benporad and Vannucchi (1963)

The closure of the Tower resulted in a strong pressure from the public for a rapid reopening, but the restoration experts warned against hasty and insufficiently pondered solutions. The Italian Government decided to install a further commission, an International Committee with the task of conceiving, designing and this time also implementing the necessary stabilization works.

The 1990 International Committee was strongly innovative because of its true interdisciplinary nature; for the first time it included not only engineers but also architects, restorers, art historians and building stones experts.[1] A geotechnical engineer, directly nominated by

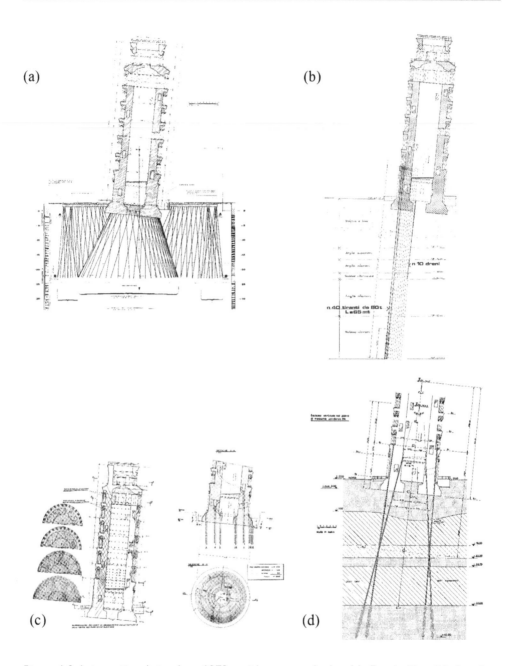

Figure 6.2 International tender, 1973, with proposals by (a) Fondedile, (b) Fondisa, (c) Geosonda and (d) Impresit Gambogi Rodio. A fifth proposal judged worthy of mention was by Konoike and consisted in consolidating the foundation soil by jet grouting.

the Prime Minister, chaired it; the other members had been designated partly by the Ministry of Public Works and partly by the newly formed Ministry of Cultural Heritage. This had the effect of a proper consideration of the safeguard of the integrity of the monument, besides its statics. Furthermore, the Committee had the characteristics of an independent authority, with adequate funding, and was free to intervene on the Tower, using a group of designers and contractors supporting it.[2]

Notes

1   The six members designated by the Ministry of Public Works were Professor G. Creazza, structural engineer; Professor J.B. Burland, geotechnical engineer; Professor M. Desideri, structural engineer; Professor Fritz Leonhardt, structural engineer; Professor G.A. Leonards, geotechnical engineer; and Professor F. Veniale, expert in building stones. The six members designated by the Ministry of Cultural Heritage were Professor R. Lemaire, art historian; Doctor M. D'Elia, restorer; Professor R. Di Stefano, restorer; Doctor M. Cordaro, art historian; Professor F. Gurrieri, restorer; and Professor C. Viggiani, geotechnical engineer. The chairman, designated by the Prime Minister, was Professor Michele Jamiolkowski, geotechnical engineer.

   During the 12 years of activity of the Committee, other experts have been members. For restoration and art history: J. Barthelemy, A.M. Mignosi Tantillo, S. Settis, A.M. Romanini; for structural engineering: R. Calzona, G. Croci, G. Macchi, L. Sanpaolesi; for geotechnical engineering: R. Lancellotta.

2   The group, called Consorzio Torre di Pisa, included an engineering firm (Bonifica SpA), three specialized contractors (Italsonda SpA, Rodio SpA, Trevi SpA) and a research center (ISMES).

Chapter 7

Leaning instability

As previously mentioned, the progressive increase of the inclination of the Tower was usually attributed to creep of the *pancone* clay. On the contrary, a careful study of the geodetic survey measurements going back to 1911 (Burland and Viggiani, 1994) revealed a most surprising form of motion of the foundation, which was radically different from previously held ideas.

The theodolite measurements showed that the first cornice had not moved horizontally, apart from two occasions in 1934 and early 1970, when anomalous movements were caused by the grouting of the foundations and by intense pumping from the lower sands. Precision leveling, which commenced in 1938, also showed that the center of the foundation plinth had not displaced vertically relative to the surrounding ground. Therefore, the rigid body motion of the Tower foundation could only be as shown in Figure 7.1, with an instantaneous center of rotation at the level of the first cornice vertically above the center of the foundation.

The direction of motion of points F_S and F_N is shown by vectors in Figure 7.1, and it is clear that the foundation has been moving with F_N rising and F_S sinking.

The discovery that the motion of the Tower was as shown in Figure 7.1 turned out to be crucial in many respects; it is probably the single most important finding in the development of the strategies for both temporary and long-term stabilization.

The form of motion is consistent with the phenomenon of "leaning instability" (Hambly, 1985), which occurs when a tall structure reaches a critical height at which the overturning moment generated by a small increase in inclination is equal to or larger than the resisting moment generated by the foundation. Leaning instability is due not to lack of strength of the ground but to insufficient stiffness.

To demonstrate leaning instability, the simple conceptual model of an inverted pendulum may be used. It is a rigid vertical pole (Fig. 7.2) with a concentrated mass W at the top and hinged at the base to a constraint that reacts to a rotation α with a stabilizing moment $M_S = \alpha/k_\alpha$; on the other hand, the rotation induces an offset of the mass and hence an overturning moment $M_O = Wh \sin \alpha$. If the stabilizing moment is larger than the overturning one, the system is stable and it returns to its initial configuration. If the contrary occurs, the system is unstable and it collapses. If the two moments are equal, the equilibrium is neutral: the system stays in the displaced configuration. The stability of the system may be characterized by the ratio $F_S = M_S/M_O$ between the stabilizing moment and the overturning one.

When the Tower is modeled as an inverted pendulum, the restraint exerted by the foundation may be evaluated by representing it as a rigid circular plate of diameter D resting on an elastic half space of constants E, v. The plate is subjected to a vertical force W applied at the height h of the center of gravity and hence with an eccentricity $e = h \sin \alpha$. The overturning

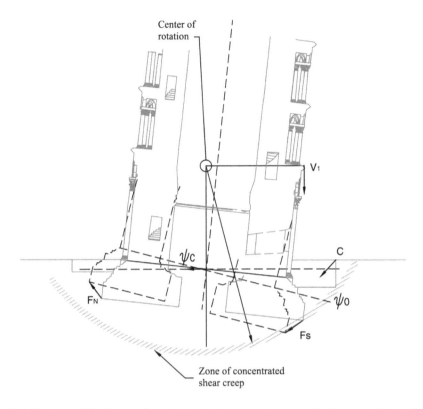

Figure 7.1 Motion of the Tower foundation during steady creep (Burland and Viggiani, 1994)

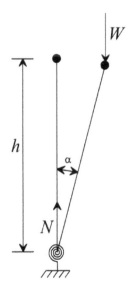

Figure 7.2 Inverted pendulum: a simple model of leaning instability

moment $M_O = W e$ and w, α are the settlement and the rotation of the foundation, respectively. It may be shown that:

$$\alpha = \frac{1}{k_\alpha} M_O \quad w = \frac{1}{k_w} W \tag{7.1}$$

with k_α and k_w given by:

$$k_\alpha = \frac{ED}{1-\upsilon^2}; \quad k_w = \frac{ED^3}{6(1-\upsilon^2)} \tag{7.2}$$

In this simple linear model, there is no coupling between settlement and rotation, and the terms k_α, k_w are intrinsic properties of the ground–monument system. The stability may be characterized by a factor of safety F_S given by the ratio between the stabilizing moment M_S and the overturning moment M_O:

$$F_S = \frac{M_S}{M_O} = \frac{k_\alpha \alpha}{Wh \sin \alpha} = \frac{ED^3}{6(1-v^2)} \frac{1}{Wh} \tag{7.3}$$

where $\sin \alpha \sim \alpha$ for the small rotations. The safety factor is also an intrinsic property of the ground–monument system.

In the case of the Tower of Pisa, an evaluation of F_S may be obtained by the knowledge of the settlement of the Tower, $w \geq 3$ m. Being $k_w = W/w$, one gets $E/(1 - v^2) \leq 2.85$ MN/m².

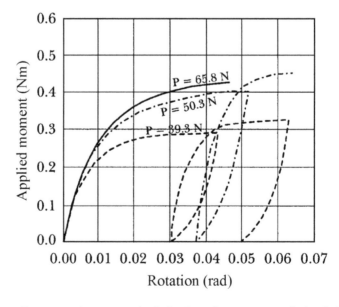

Figure 7.3 Centrifuge experiments on the behavior of an eccentrically loaded circular rigid plate resting on a clay subsoil (Cheney et al., 1991)

Accordingly, with $h = 22.6$ m (height of the center of gravity of the Tower) and $W = 141.8$ MN (weight of the Tower), $F_s \leq 1.12$. Even the simplistic linearly elastic subsoil model allows the important conclusion that the Tower is very near to a state of neutral equilibrium. Detailed studies revealed that the continuing movement, made possible by the state of neutral equilibrium, is controlled by ratcheting due to seasonal cyclic actions of the fluctuating water table in Layer A. Of course, creep must have also some influence on the process.

The relationship between the stabilizing moment M_s and the rotation α may be linearized over a short interval, but it is certainly nonlinear and asymptotically approaches a limiting value of M_s. In a case such as that of the Leaning Tower, which is on the verge of instability, consideration of nonlinearity appears mandatory. Centrifuge experiments by Cheney et al. (1991) (Fig. 7.3) show, in fact, nonlinearity, strain-hardening plasticity and coupling between vertical force and rotation. Referring again to the inverted pendulum model, in the case of nonlinearity and strain-hardening plasticity, the relation between loads and displacements has to be expressed in incremental form. Equation (7.1) becomes:

$$d\alpha = \frac{\partial M}{k_\alpha} + \frac{\partial W}{k_{\alpha w}}; \quad dw = \frac{\partial M}{k_{w\alpha}} + \frac{\partial W}{k_w} \tag{7.4}$$

The k_{ij} terms depend on the load increment, the current state of load and the load history. Hence the factor of safety depends on the current state of stress and stress history.

It may be seen in Figure 7.3 that a decrease of the inclination, bringing on the unloading branch of the curve, strongly increases the stiffness k_α of the ground–foundation system (represented in the diagram by the tangent to the $M_s - \alpha$ curve) and hence the stability. This generated the idea that a decrease of inclination could be used to stabilize the Tower.

Chapter 8

The work of the International Committee, 1990–2001

8.1 Temporary stabilization

Fully aware that the conception, design and implementation of the permanent stabilization measures require time, the Committee took an early decision to implement temporary and fully reversible interventions to slightly improve the safety against overturning and to gain the time needed to properly devise, design and implement the permanent solution (Burland *et al.*, 2000).

Since it had been observed that the north side of the Tower foundation had lifted throughout the 20th century, it was felt that application of a vertical force at that point would have been efficacious in reducing the overturning moment and in improving the conditions of stability in the monument. A solution of this type is conceivable only if the trouble with the Tower is seen as a phenomenon of leaning instability; it would never have been adopted if the problem had been that of a bearing capacity failure.

Before beginning, the intervention scheme was studied by means of in-depth analyses to ascertain its safety and that it would not give rise to any undesired effects. Another aim of the study was to forecast the Tower's response to provide a prediction with which to compare later the data emerging from observations.

The analyses (Burland and Potts, 1994), performed with a finite-element model based on a nonlinear elastic isotropically hardening plastic constitutive model of the subsoil, showed that it would have been possible to apply safely a maximum load of 14 MN. Beyond that threshold, there was the risk of inducing plastic deformations in the *pancone* clays, with extensive settlement and rotation toward the south.

The plan for application of a counterweight on the north side was developed taking into account these results (Viggiani and Squeglia, 2005c; Viggiani *et al.*, 2005b). A prestressed reinforced concrete ring beam was placed around the base of the Tower at the level of the steps that emerge from the bottom of the *catino* and was used as the base for a stack of lead ingots weighing about 100 kN each. The total load – equal to about 6 MN, including the weight of the ring beam – was applied gradually between May 1993 and January 1994 (Fig. 8.1).

The counterweight induced a change of inclination of 33 arc seconds by February 1994; by the end of July, it had increased to 48 arc seconds and eventually to 52 arc seconds. On February 1994, the average additional settlement of the Tower relative to the surrounding ground was about 2.5 mm. An event of the utmost importance is that the progressive southward inclination of the Tower had come to a standstill.

Figure 8.1 Temporary intervention by lead ingot counterweights

These observations allow an important experimental confirmation of the intended stabilization approach. The Tower reacted to the application of the lead weights with a rotational stiffness:

$$k_\alpha = \frac{43.5 \text{ MNm}}{52 \text{ arc seconds}} = \frac{43.5 \, MNm}{2.51 \times 10^{-4} \, rad} = 172,548 \text{ MNm} \qquad (8.1)$$

The factor of safety has been increased to:

$$F_s = \frac{k_\alpha}{Wh} = \frac{172,548}{141.8 \times 22.6} = 54 \qquad (8.2)$$

As a matter of fact, after the application of the counterweight, the Tower remained essentially motionless for over three years, apart from seasonal cyclic movements.

At this point, it must be mentioned that the legislative decree that established the Committee expired many times, but only occasionally was it ratified by Parliament in time; for this reason, it had to be reinstated by the executive branch of government. Among other things, over the course of its troubled lifetime, the Committee lost its mandate a number of times for longer or shorter periods. Shortly after the positive experience with the lead counterweights, there began to arise concern that the Committee could be disbanded entirely (as had happened in the cases of the earlier Commissions) and the unsightly lead weights be left in place on the Tower for an indefinite but certainly very long period of time.

8.2 The 10-anchor solution

With this in mind, an alternative system to the counterweights was conceived (Viggiani and Squeglia, 2005d), based on the use of 10 ground anchors consisting in steel cables cemented into the lower sands at a depth of 45 m, as shown in the schematic diagram in Figure 8.2. The lead ingots would have been removed progressively as the anchors were tensioned. With the forces applied being equal, this solution was more advantageous than the counterweights from the static point of view, thanks to the greater lever arm (9.6 m instead of the 6.5 m of the ingots). By varying the tensioning of the single anchor, it would have been possible to "guide" the movements of the Tower. The solution was in any case conceived as a temporary action, although with a medium-term duration (some decades).

The major difficulty with this solution was the need to create a new prestressed ring beam to transfer the force from the anchors to the base of the Tower. In order to be invisible from the outside, this beam had to be built below the floor of the *catino*; for this reason, it was necessary to excavate around the base of the Tower to below the groundwater level. With the effects of the excavation of the *catino* carried out in 1838 fresh in mind, the Committee

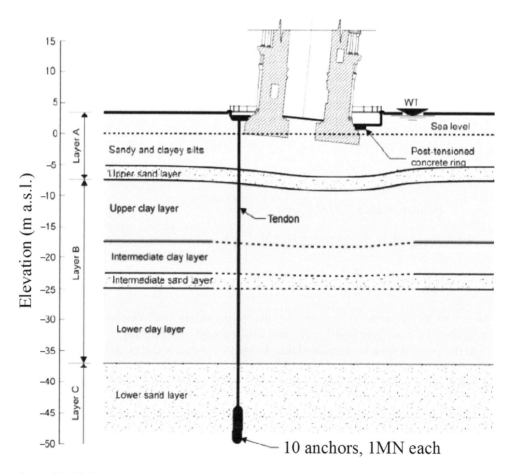

Figure 8.2 Medium-term solution of 10 anchors (not implemented)

was fully aware of the risks involved. It was thus decided to freeze the ground surrounding and underneath the excavation site. The ground to be frozen, however, was well above the Tower's foundation plane so as to exclude the Tower from any movement caused by freezing. The freezing plant consisted of a series of steel pipes inserted into the ground, in which liquid nitrogen was circulated.

Following investigations by drill cores, it was discovered that below the *catino* floor is a concrete bed of about 1-m thickness, set in place partly in 1837 and partly in 1935. The finding of a circumferential gap at the interface between the concrete and the Tower foundation led to the conclusion that the two bodies were not connected; therefore, the volume variations of the frozen soil during freezing and thawing were not expected to influence the Tower.

Without going into the details of the operation, it can be reported that freezing was commenced on the north side, and the northern sections of the ring beam were successfully installed. They were connected to the foundation by means of stainless steel rods, cemented into the foundation masonry. During these operations, the water tightness of the *catino* was partly destroyed, and two pumps had to be installed to prevent the flooding; since a sand layer was provided below the ring beam sections, the system worked as a groundwater level control.

During the excavation, short steel tubes of the external diameter of about 60 mm emerging from the Tower foundation were discovered. They were the intakes of the injection holes executed in 1934–1935 to make the foundation watertight and had been left in place. Despite their short length (about 0.5 m, half cemented in the masonry and half protruding), they provided some connection between the concrete bed below the *catino* floor and the Tower.

September 1995 (in the Committee's words, "Black September") saw the beginning of freezing operations southeast and southwest of the Tower. The Tower began to rotate to the south at a fast rate; the movement was also affected by an attempted installation of some micropiles at the south boundary of the *catino*. Freezing was interrupted, and the rotation was controlled by adding further lead ingots on the north side. By the time the implementation of the 10 anchors solution was abandoned, the residual increment in inclination to the south was only 7 arc seconds, but the counterweight had exceeded 10 MN. Again, good intentions did not produce good results!

Figure 8.3 shows the variations in inclination obtained with the lead counterweights (1993–1994) and the perturbing effects of the attempt to implement the 10 anchors solution (September 1995).

8.3 Final stabilization

The Committee had developed deep insight into the behavior of the Tower through the interpretation of its history, scrutiny of the measurements taken in the last century and analysis of the phenomenon of leaning instability. After a comprehensive discussion, the Committee concluded that a decrease of the inclination of the Tower by half a degree (1800 arc seconds, that is, around 10% of the inclination in 1990) would have been sufficient to stop the progressive increase of inclination and would substantially improve stability conditions. At the same time, such a reduction was considered small enough not to be perceived at first sight.

The decrease had to be obtained by inducing a differential settlement of the Tower opposite to the existing one and hence by acting on the foundation soil and not on the Tower. This scheme would also cause a reduction of the stress in the masonry on the south side

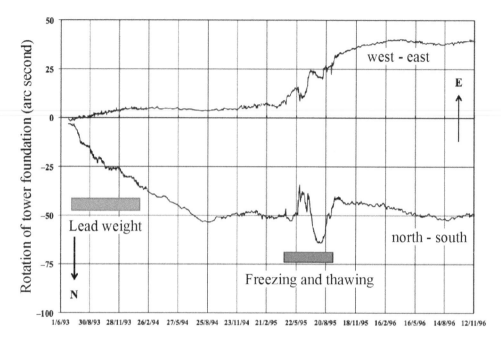

Figure 8.3 Rotations caused by the lead counterweights and by the freezing and excavation works

and therefore would minimize the required structural reinforcement; moreover, it was fully respectful of the formal, as well as the material and historical, integrity of the monument. As Salvatore Settis so aptly observed, "[A] variation in inclination was already dictated by the Tower's genetic code."

The Committee examined different possible means for achieving the required reduction in inclination of the Tower. Among these, for example, were the creation of a reinforced concrete slab on the ground surface north of the Tower, so as to apply a load to the ground through pretensioned steel cables anchored in the lower sands. Another was the shrinking of the upper clays, again north of the Tower, by means of vacuum pumping or electroosmosis (Esrig, 1968). All these solutions were studied by means of numerical analyses, on small-scale models at natural gravity and in a centrifuge, and with large-scale *in situ* trials (Viggiani and Squeglia, 2003). Following these studies and investigations, it was finally decided to induce settlement of the north side of the Tower by controlled extraction of small volumes of soil to the north of and from below the foundation level (underexcavation).

The decision was not easy. The Committee had to consider some doubts within the Committee itself and two strong opposing pressures from the public.

The first may be exemplified by the following episode. Fernando Lizzi (1914–2003) was a very bright Neapolitan engineer, considered the father of micropiles technology. In a paper presented to an international symposium (Lizzi, 2000), he recalls that he had submitted to the 1973 international tender a project based on the use of micropiles (proposal (a) in Fig. 6.2). The project was signed by the same Lizzi and Jean Kerisel: names that are a guarantee of quality.

Lizzi (*loc. cit.*) writes:

> The Competition was not awarded; no decision was taken and, at the date of the present paper, the problem is still in the hands of a special Committee, appointed ten years ago. As for the above project, based on a network of Pali Radice, the present Committee admits its full validity from the engineering point of view; but its members solemnly declare that it cannot be accepted because the execution of piles, although concealed in the low masonry and in the subsoil . . . *spoils the integrity of the Monument.* . . . Therefore, the Committee is looking for a solution which can be carried out *without touching* the Monument.

Lizzi makes sarcastic references to the Committee and a large majority of civil engineers would probably agree with his position: the common sense of a familiar, good, reliable underpinning to be obviously preferred to the apparently meaningless pretension of stabilizing the Tower without even touching it!

The second external pressure is subtler but is essentially based on the firm *a priori* belief that the Tower cannot be studied simply by the laws of mechanics, since it is a kind of living organism, probably capable of finding its internal stability by itself. This position, for instance, is expressed in a very enjoyable book by Pierotti (1990).

Underexcavation for the stabilization of the Leaning Tower of Pisa had been suggested by Ferdinando Terracina (1962). Singularly appropriate historical precedents were the 19th-century interventions on leaning bell towers performed in England and in Holland and an intervention on an industrial smokestack in Germany (Johnston and Burland, 2004; Burland and Johnston, 2005). The method had been successfully applied more recently in Mexico; among others, a particularly important application of the technique was that aiming at mitigating the effects of the highly visible differential settlements of the Metropolitan Cathedral of Mexico City (Ovando-Shelley and Santoyo, 2013).

Underexcavation consists in extracting a small volume of soil in a carefully chosen position, leaving in its place a cavity. Under the action of the pressure exerted on the ground, the cavity closes progressively and induces a limited settlement at the surface level. The process is repeated at various points so that the surface settlement proceeds gradually and in an easily controllable manner.

Small-scale model tests of underexcavation at natural gravity and in the centrifuge, in addition to numerical analyses, gave a favorable response, encouraging the Committee to undertake a large-scale experiment to develop the field equipment and to explore the operational procedures. For this purpose, a 7-m-diameter eccentrically loaded instrumented footing was constructed in the Piazza and subjected to underexcavation (Burland *et al.*, 2000) (Fig. 8.4).

The trial was very successful; it proved that it was possible to steer the footing and act on the rate of rotation by varying the position and the intensity of soil extraction. The numerical modeling and the centrifuge tests had revealed the existence of a critical line beyond which the effect of the underexcavation is detrimental; in fact, during the early stages of the field trial, an overenthusiastic excavation beyond that line produced an increment of the inclination thus confirming the prediction.

Being aware that the investigations carried out might be not completely representative of the possible response of a tower affected by leaning instability, the Committee decided to implement a preliminary and limited ground extraction beneath the Tower itself in order

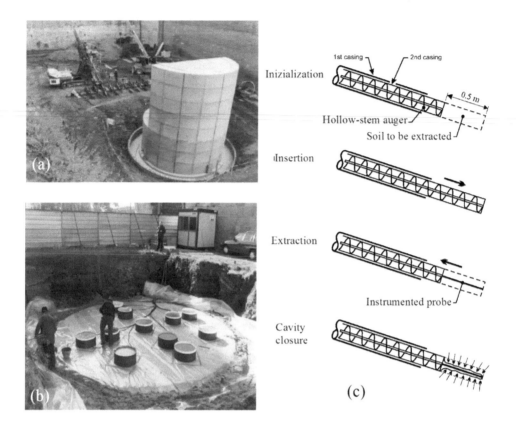

Figure 8.4 (a) Underexcavation trial field; (b) installation of pressure cells on the foundation plane; (c) underexcavation procedure finally developed

to observe its response. To prevent any unexpected adverse movement of the monument, a safeguard structure was necessary. It consisted of two sub-horizontal steel stays (Fig. 8.5), connected to the Tower at the level of the 3rd order and to two frames located some 100 m apart; it was capable of applying to the Tower a stabilizing moment but only if needed. The safeguard structure was installed in December 1998.

The preliminary underexcavation experiment was carried out between February and June 1999, operating with 12 inclined drill holes and removing a total of 7 m³ of soil, of which 71% was north of the Tower and 29% was from beneath the foundation. The extraction of soil was very gradual, with the removal of only 22 dm³ in each single extraction. The movements of the Tower were monitored by a comprehensive system of geodetic survey and electronic tilt meters, allowing the acquisition and processing of the data in real time. An observational control approach was adopted; the program of soil extraction was decided daily, based on the observed movements of the previous day.

The Tower rotated northward by 90 arc seconds until June 1999, when the preliminary soil extraction operation ceased; by mid-September, the rotation had increased to 130 arc

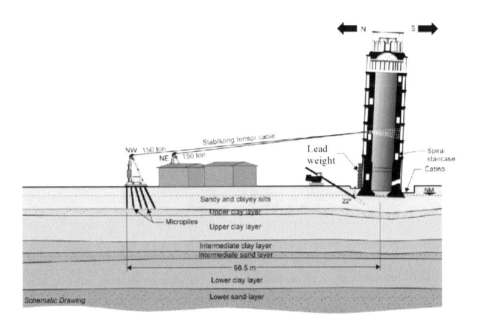

Figure 8.5 Scheme of the safeguard structure with steel stays

seconds. At that time, three of the 97 lead ingots were removed, and the Tower then exhibited negligible further movements.

After the very positive results of the preliminary underexcavation, the Committee agreed to proceed forthwith to the full underexcavation (Fig. 8.6). It was carried out between February 21, 2000 and June 6, 2001, with 41 holes, removing a total of 38 m³ of soil (70% below the *catino*, that is, outside the perimeter of the foundation) in a total of 1737 single extractions. During the same period, all the lead ingots were removed. In June 2001, the steel safeguard cable stays were dismantled, without having ever been operated. The goal of reducing the inclination of the Tower by half a degree had been fully attained (Fig. 8.7). We saw previously that Figure 4.3 depicted the deduced history of inclination of the Tower up to 1993. Figure 8.8 completes the history up to the time of writing and shows that the works described in this book have had the effect of bringing the Tower back to the inclination it had prior to Gheradesca's near disastrous excavation of the *catino* in 1838.

8.4 Additional stabilizing measures

It is to be added that the study of the movements of the Tower, depicted previously in Figure 7.1, led to the important conclusion that the seat of the continuing long-term rotation of the Tower lies in Layer A and not within the underlying clay, as had been widely assumed in the past. Consistent with the seat of the movement lying within Layer A, it was concluded from a close study of the seasonal movements of the Tower and the levels of the water table that, in addition to creep, the most likely cause of the progressive rotation was a fluctuating groundwater level due to heavy rainstorms.

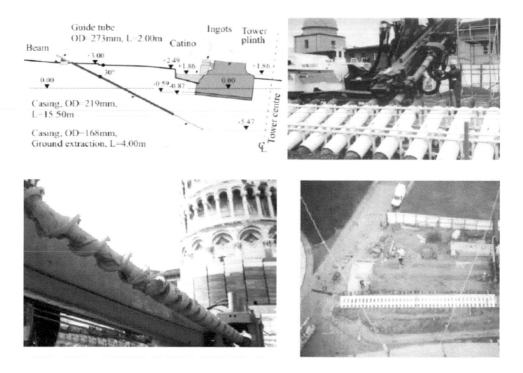

Figure 8.6 Final underexcavation

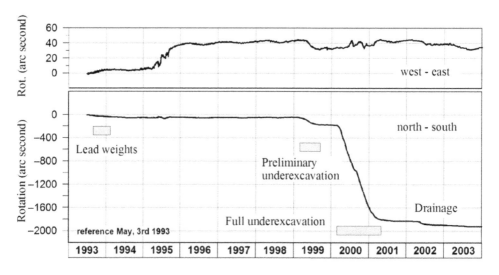

Figure 8.7 Rotation of the foundation of the Tower since 1993

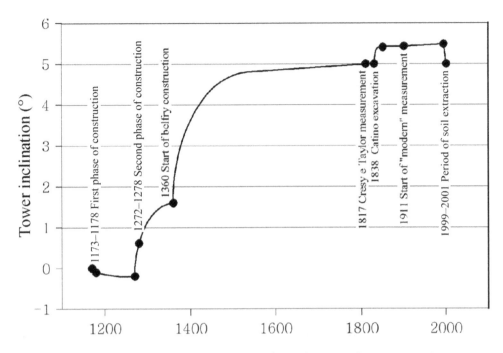

Figure 8.8 History of the rotation of the Tower from the time of its construction

Piezometric measurements made over several years have confirmed this hypothesis. In fact, the average groundwater level close to the south side of the Tower in Layer A is 200–300 mm higher than that to the north, as shown in Figure 8.9. This difference generates a small but not negligible stabilizing moment for the monument that is so close to falling over. In the autumn and winter, when the rainfall events are more intense, the water table rises sharply, reducing the difference in piezometric levels and thereby producing southward rotations of the Tower, which are not fully recovered. It is believed that the cumulative effects by ratcheting of these repeated impulses has been one of the factors producing the steady increase of inclination in the long term. To minimize this effect, it was necessary to eliminate the fluctuations of the water table, and, with this objective in mind, a drainage system was installed consisting of three wells sunk on the north side with radial subhorizontal drains running into them from below the north side of the *catino* (Fig. 8.10). The water level in the wells is controlled by the level of the outlet pipe. The drainage system was implemented in April and May 2002 and led to a decrease in the pore water pressure, as well as a significant reduction in its seasonal fluctuation, as shown in Figure 8.11. The installation of this drainage system, furthermore, induced another northward rotation of the Tower, as can be seen in Figure 8.7.

In addition to reducing the inclination of the Tower by half a degree and controlling the fluctuations of the groundwater, another contribution to the foundation stabilization has been provided. As previously mentioned, during the work of the Committee, a 0.8-m-thick cement conglomerate ring was detected in the bottom of the *catino* around the base of the Tower. This ancient concrete ring is of high quality; it has now been connected to the masonry

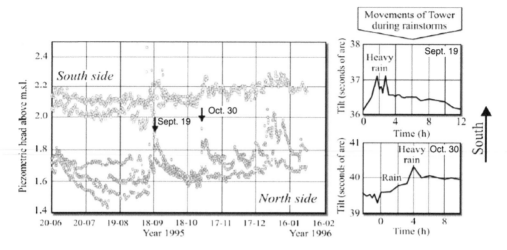

Figure 8.9 Groundwater fluctuation in Layer A

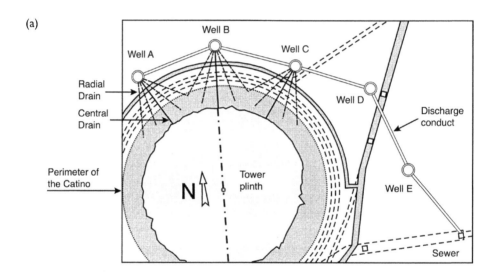

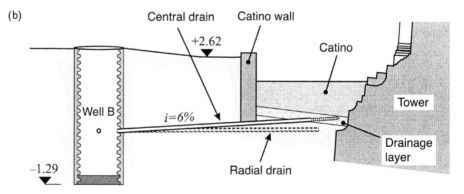

Figure 8.10 Drainage system to control ground water level in Layer A, on north side: (a) plan view; (b) cross section

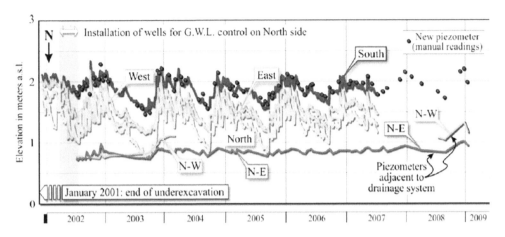

Figure 8.11 Groundwater level in Layer A after the implementation of the drainage system

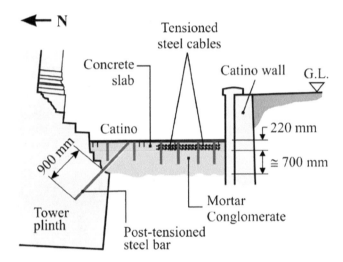

Figure 8.12 Structural connection of the concrete ring below the floor of the *catino* to the Tower foundation

foundation of the Tower by means of stainless steel reinforcements and has been strengthened by circumferential post tensioning (Fig. 8.12). As a result, the effective area of the foundation has been increased, thereby further increasing the factor of safety against leaning instability. The Tower was reopened to the public by the Minister of Works on June 16, 2001 (Fig. 8.13), very appropriately the day before the feast of San Ranieri, the patron saint of Pisa whose funeral cortege is depicted in Figure 4.4.

Figure 8.13 Reopening of the Tower to visitors on June 16, 2001

Chapter 9

And now?

9.1 Behavior of the Tower after stabilization

At the time of writing, we have at our disposal a monitoring period of some 19 years after the stabilization work. Figure 9.1 reports the observed changes in inclination of the Tower.

At a first glance, the situation appears satisfactory; the Tower is still slowly moving northward and approaching a motionless state with a decreasing rate. Soon after the stabilization measures, the rate of rotation was decreasing rapidly: 9 arc seconds per year in 2004, 7 arc seconds per year in 2005, 6 arc seconds per year in 2006. Then the decrease became slower: 2 arc seconds per year in 2007; 1 arc second per year in 2018. In any case, the honeymoon is continuing.

As for the seasonal movements, Figure 9.2 compares those of the period before stabilization, already presented in Figure 5.4, with those after the stabilization. The comparison has some limits. First of all, the observation period of the two series of measurements is very different: 16 years for the latter, 55 years for the former. Second, measurements refer to the same rotation of the whole Tower but measured with different instruments: the GB pendulum

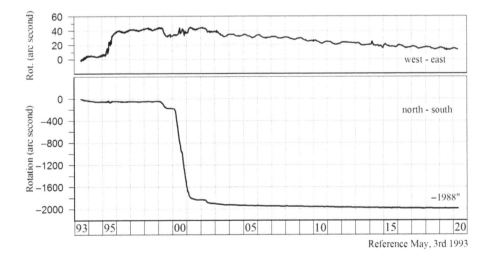

Figure 9.1 Rotation of the Tower till March 2019

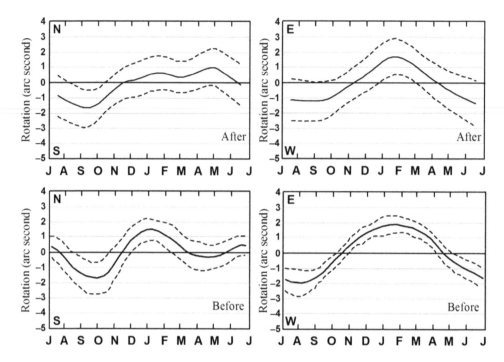

Figure 9.2 Comparison of the seasonal movements before and after the stabilization

before the underexcavation, an electronic sensor after it. As a consequence, the old measurements were taken once each day, around 9 a.m. whereas the new measurements are taken at one-hour time intervals, and therefore they are influenced by temperature variations during the day. Finally, the climatic differences between the two periods and the regulation of the groundwater table may have an influence. In any case, Figure 9.2 suggests that, after the stabilization intervention, seasonal movements are similar to the previous ones but with slightly smaller amplitudes both in the north–south and the east–west directions. This is certainly a reassuring observation.

At this point, however, the reader will probably ask the question: how will the Tower behave in the future? Attempting an answer is not easy due to the complexity of the phenomena involved and the number and variety of factors influencing them. In any case, we don't want to escape the question; our tentative answers are the subject of the next section.

9.2 Future scenarios: extrapolation

Upon concluding its work, the International Committee outlined two possible scenarios (Jamiolkowski, 2005).

In the first one, rather conservative and perhaps pessimistic, the Tower will remain motionless for some decades (a period that the Committee called the "honeymoon") and then gradually resume a southward rotation, first at a rather slow rate and then progressively accelerating. In this scenario, should the rate of rotation be the same as before the intervention, the Tower would reach the value of the inclination it had in 1993 in a time span in the order of three centuries. Should better options not become available, before reaching this point the underexcavation intervention could be repeated.

In a more optimistic and perhaps more probable scenario, the Tower will gradually attain a motionless situation, apart from the cyclic movements caused by daily sun irradiation, seasonal changes in the water table and the influence of the generalized subsidence of the whole Pisa plain, which affects the Piazza and the Tower (Croce *et al.*, 1981).

How to choose between the two? On a purely empirical basis, we can attempt to extrapolate the available observations. Alternatively, we can base our predictions on some analysis, using a numerical model validated against the observed behavior.

Two possible extrapolations into the future decades are reported in Figure 9.3, in which the precision leveling of the eight internal points have been used. The red line is a parabola, the green line a hyperbola. The parabola has a point of maximum, with a cruel end of the honeymoon on May 30, 2027, followed by an inversion of the motion; thus it may be representative of the preceding first scenario. The hyperbola tends to an asymptote of 2,160 arc seconds; it may be representative of the second, more optimistic scenario.

What is very interesting to note is that, from a statistical viewpoint, the two curves are practically equivalent; this means that, with the available data, it is not possible to foresee the scenario that will actually occur in the long term. It can be added that many attempts to improve the reliability of the extrapolation, such as adopting different curves, have not modified this conclusion.

The only indication that can be obtained by the mere extrapolation of the available data is that in the next decade the Tower will keep rotating toward north. In 10–15 years' time, it will be possible to understand whether the Tower will reverse the sense of rotation and start rotating again to the south or go on approaching a motionless state.

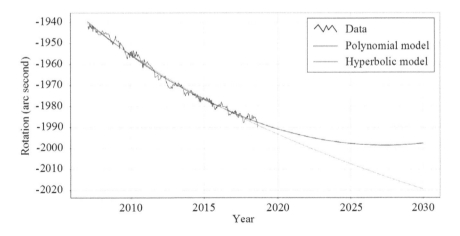

Figure 9.3 Two possible extrapolations of the inclination measurements

9.3 Future scenarios: numerical analysis

A first attempt to describe by numerical analysis the time–deformation history of the Tower was carried out in 2002 (Vermeer *et al.*, 2002). A mesh of 4400 elements with six integration points was employed, assuming symmetry about the north–south vertical plane because of the limited computational power at the time. The clayey and silty layers were modeled by the Soft Soil Creep (SSC) model, while the other layers were considered simple elastic–perfectly plastic materials. A second, improved analysis was carried out later (Leoni and Vermeer, 2010), still assuming symmetry and employing a mesh of 9500 elements and with a more detailed and realistic modeling of the geometry and a better simulation of the history. For the topmost layers, the Hardening Soil Small Strain model was adopted.

A third step of the analysis has been carried out recently (Wesi Geotecnica, 2016), taking advantage of the greater computational power available. The full geometry of the Tower and its subsoil has been considered, with a finite element (FE) mesh consisting of 230,000 tetrahedral elements. Furthermore, a new original numerical procedure was developed to carefully simulate micropiles construction and underexcavation (Squeglia *et al.*, 2018).

The full results of the analysis will be presented and discussed elsewhere. We observe just that the improved analysis has been rather successful in describing some parts of the recent history of the Tower; Figures 9.4–9.6, for instance, compare the results obtained by the analysis to the observed behavior during micropiles installation in 1995, final underexcavation in 1999/2001 and groundwater regulation in 2002/2003.

The analysis, on the contrary, has substantially overestimated the rotations generated by the lead counterweight.

For the purpose of predicting the future, a comparison between computed and observed behavior of the Tower during and after the stabilization works, in terms of rotation and settlement, is reported in Figures. 9.7 and 9.8.

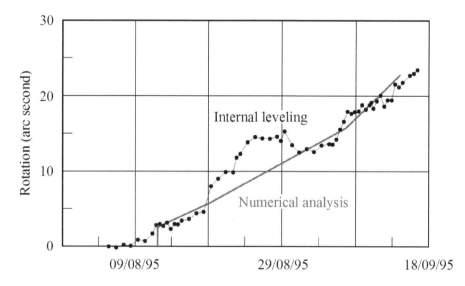

Figure 9.4 Measured versus computed rotation due to micropiles installation

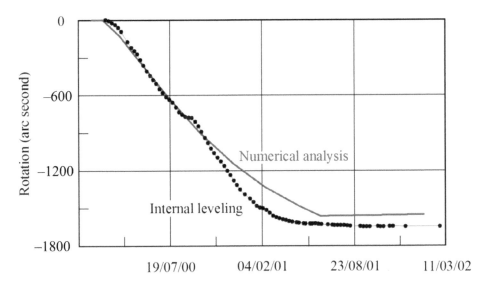

Figure 9.5 Calculated versus measured rotation after the final underexcavation

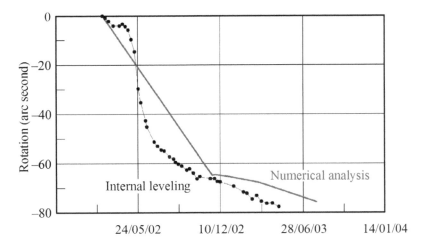

Figure 9.6 Calculated versus measured rotation after the groundwater regulation

The behavior of the Tower after the stabilization work, as described by the monitoring, is characterized by a progressive increase of the average settlement, at a rate of about 0.5 mm per year, and a progressive rotation toward north, as discussed in Section 9.1.

The numerical analysis is rather successful <u>as</u> far as settlement is concerned (Fig. 9.8). On the contrary, however, the prediction of rotation increase (Fig. 9.7) does not agree with the available observations. The analysis predicts a steady rotation <u>southward</u> at a constant rate of about 3 arc seconds per year over the 19 years of measurements, whereas a <u>northward</u> rotation has been observed.

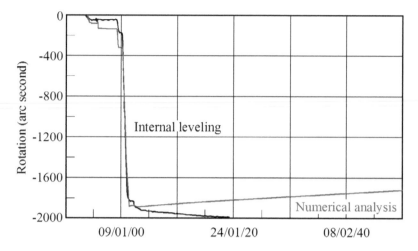

Figure 9.7 Calculated versus measured Tower rotation since 1993

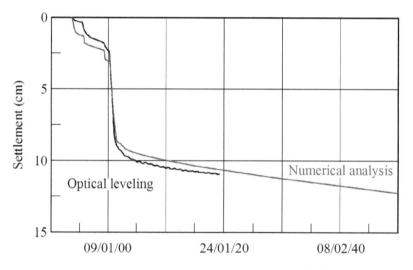

Figure 9.8 Calculated versus measured Tower settlement since 1993

It could be argued that the available observations concern a transient stage and that in the long term the rotation will change sign; again, we must wait some 10–15 years to verify this hypothesis.

Both the extrapolation and the numerical analysis lay stress on the importance of continuing the monitoring of the Tower and the interpretation of its results.

Chapter 10

Concluding remarks

Many lessons may be learned from an experience like that of the intervention to stabilize the Leaning Tower of Pisa.

The ancient masons, the *magistri lapidum*, took two centuries to construct the Tower, persevering in completing it despite many difficulties and an evidently increasing inclination. The achievement of the International Committee, for its part, should be seen as the culmination of over a century of studies, researches, errors and attempts, pursuing a difficult balance between the requirements of statics and integrity. The first lesson, hence, is that of persevering: the cooperation among different disciplines, the careful comprehension of the history and a study of the behavior of the monument allowed a result that was anything but obvious.

Such a result has been attained also because the enabling Parliamentary act gave the Committee the role of autonomous authority, allowing it to carry out studies, investigations and interventions free from the usual bureaucratic constraints, taking advantage of the Consorzio, its operational arm. Bureaucratic constraints (Italy has unquestionably a leading position in this field) can override the best of endeavors, and they were partially encountered in the initial stage of the Committee's life.

A third lesson may be expressed with the proverb "The road to hell is paved with good intentions." In the mid-19th century, the architect Gherardesca, moved by the desire to save the heritage of the ancestral greatness, excavated the *catino*; this has been extremely detrimental to the Tower's safety. The intervention to make the *catino* watertight in the 1930s and the attempt of the Committee to implement a provisional intervention with ground anchors also had detrimental effects on the Tower. Good intentions are certainly necessary but are not enough on their own.

Two historical studies proved to be most valuable in arriving at suitable stabilization measures. The first was a study of the history of inclination of the Tower during and subsequent to construction. This study was used to calibrate a sophisticated numerical model of the Tower and the underlying ground. Reliance was placed in this model for evaluating the effectiveness of various stabilization schemes. The second study was of measurements of movement made since 1911. This latter study revealed an unexpected mechanism of foundation movement which proved crucial in developing both temporary and permanent stabilization measures.

It is known that most major undertakings rest, at least partially, on the effects of casual favorable circumstances. The Tower, throughout its long history, has been protected by a benevolent fate. The two long interruptions during construction that saved it from an early collapse; the shortsightedness of Sergeant Weckstein; the avoidance of invasive and maybe

Figure 10.1 Proud and mysterious stranger

dangerous interventions during the second half of 20th century – are all events confirming the intervention of fate. Another important lesson, therefore, is that a modicum of good luck won't do any harm. This is a very welcome conclusion for the two Neapolitan authors, but in this connection it is worth recalling Skempton's (1961) wise assertion that "optimism and over-confidence may impress one's client, but they have no influence on the great forces of nature."

In the end, however, let us recall those days spent in Piazza dei Miracoli, gazing up at a monument in the dreamworld colors of something a few meters and many centuries away: the apse of the Cathedral, with at its top the bronze griffon (Fig. 10.1).

The griffon. The proud and mysterious stranger came one day from the Orient, over that sea that today shimmers at the horizon behind it. And it continues to turn its enigmatic, remote gaze over the crowds in that piazza that it has been regarding unperturbed since the 11th century.

And we will never forget it.

References

Ammannati, G. (2018) La firma ritrovata: Bonanno e la Torre di Pisa. *Annali della Scuola Normale Superiore di Pisa, Classe di Lettere e Filosofia*, serie 5, 10(2), 383–398.

Burland, J.B. & Johnston, G. (2005) Historic examples of underexcavation. *Bollettino d'Arte, Vol. Spec. 2005: La Torre Restituita*, 1, 255–264.

Burland, J.B. & Potts, D.M. (1994) Development and application of a numerical model for the leaning tower of Pisa. In: *IS Prefailure Deformation Characteristics of Geomaterials*. Balkema, Rotterdam, Vol. 2. pp. 715–738.

Burland, J.B. & Viggiani, C. (1994) Osservazioni del comportamento della Torre di Pisa. *Rivista Italiana di Geotecnica*, 28(3), 179–200.

Burland, J.B., Jamiolkowski, M.B. & Viggiani, C. (2000) Underexcavating the tower of Pisa: Back to the future. In: Balasubramanian, A.S. *et al.* (eds) *Geotech Year 2000, Development in Geotechnical Engineering*. Asian Institute of Technology, Bangkok. pp. 273–282.

Cheney, J.A., Abghari, A. & Kutter, B.L. (1991) Leaning instability of tall structures. *Journal of Geotechnical Engineering, Proc. ASCE*, CXVII, 2, 297–318.

Cresy, E. & Taylor, G.L. (1829) *Architecture of the Middle Ages in Italy, Illustrated by Views, Plans, Elevation, Sections and Details of the Cathedral, Baptistery, Leaning Tower of Campanile and Campo Santo at Pisa from Drawings and Measurements Taken in the Year 1817*. Published by the Authors, London.

Croce, A., Burghignoli, A., Calabresi, G., Evangelista, A. & Viggiani, C. (1981) The tower of Pisa and the surrounding square: Recent observations. *X International Conference on Soil Mechanics and Foundation Engineering*. Balkema, Rotterdam, Vol. III. pp. 61–70.

Esrig, M.I. (1968) Pore pressure, consolidation, and electrokinetics. *Journal of the Soil Mechanics and Foundations Division, Proceedings of ASCE*, 94(SM4), 899–921.

Franchi Viceré, L., Viggiani, C. & Squeglia, N. (2005) La Piazza del Duomo: Sottosuolo, archeologia, storia. *Bollettino d'Arte, Vol. Spec. 2005: La Torre Restituita*, 1, 11–42.

Franchi Viceré, L., Veniale, F., Lodola, S. & Pepe, M. (2005) La Torre campanaria. *Bollettino d'Arte, Vol. Spec. 2005: La Torre Restituita*, 1, 43–88.

Hambly, E.C. (1985) Soil buckling and leaning instability of tall structures. *The Structural Engineer*, 63A(3), 77–85.

Jamiolkowski, M.B. (2005) Introduzione. *Bollettino d'Arte, Vol. Spec. 2005: La Torre Restituita*, 1, 1–7.

Johnston, G. & Burland, J.B. (2004) Some historic examples of underexcavation. In: Jardine, R.J., Potts, D.M. & Higgins, K.G. (eds) *International Conference on Advances in Geotechnical Engineering; the Skempton Conference*. Thomas Telford, London, Vol. 2. pp. 1068–1079.

Leoni, M. & Vermeer, P.A. (2010) *3D creep analysis of the leaning tower of Pisa*. Unpublished research report, Stuttgart University.

Leoni, M., Squeglia, N. & Viggiani, C. (2018) Tower of Pisa: Lessons learned by observations and analysis. In: Lancellotta, R., Flora, A. & Viggiani, C. (eds) *Geotechnics and Heritage: Historic Towers*. Taylor and Francis, London. pp. 15–38.

Lizzi, F. (2000) Micropiles: Past, present and future. In: Balasubramanian, A.S. *et al.* (eds) *Geotech Year 2000, Development in Geotechnical Engineering*. Asian Institute of Technology, Bangkok. pp. 145–152.

Locatelli, P., Polvani, G. & Selleri, F. (1971) Caratteristiche geometriche e fisiche della Torre e suo stato di conservazione. In: *Ricerche e studi sulla Torre pendente di Pisa ed i fenomeni connessi alle condizioni d'ambiente*. IGM, Firenze, Vol. 1. pp. 17–68.

Lumini, U. & Polvani, G. (1971) Indagini storiche e strutturali sulla Torre in relazione al suo dissesto rispetto alla verticale. In: *Ricerche e studi sulla Torre pendente di Pisa ed i fenomeni connessi alle condizioni d'ambiente*. IGM, Firenze, Vol. 1. pp. 69–96.

Macchi, G. & Ghelfi, S. (2005) Analisi strutturali. *Bollettino d'Arte, Vol. Spec. 2005: La Torre Restituita*, 3, 73–130.

Ministero LL.PP. (1971) *Ricerche e studi sulla Torre pendente di Pisa ed i fenomeni connessi alle condizioni d'ambiente*. IGM, Firenze, 3 Vol.

Ministero P.I., Opera Primaziale. (1913) *Relazioni compilate dalla Commissione Tecnica per lo studio delle condizioni presenti del campanile di Pisa*. Tipografia Galileiana, Firenze. p. 159.

Noccioli, R., Polvani, G. & Salvioni, G. (1971) I movimenti della torre dal giugno 1911 a tutto il 1968. In: *Ricerche e studi sulla Torre pendente di Pisa ed i fenomeni connessi alle condizioni d'ambiente*. IGM, Firenze, Vol. 1. pp. 97–150.

Ovando-Shelley, E. & Santoyo, E. (2013) Contribution of geotechnical engineering for the preservation of the Metropolitan Cathedral and the Sagrario Church in Mexico City. In: Bilotta, E., Flora, A., Lirer, S. & Viggiani, C. (eds) *Geotechnics and Heritage*. Taylor and Francis, London. pp. 153–178.

Pierotti, P. (1990) *Una torre da non salvare. Storia di 16 commissioni, 147 commissari, 150 anni di progetti per la Torre di Pisa. Probabilitas (1992) Statistical Analysis of the Rotation*. Unpublished report, Opera della Primaziale Pisana.

Rohault, de Fleury, Charles. (1859) Le Campanile de Pise. In: *Encyclopédie de l'architecture*, Bance, Paris.

Skempton, A.W. (1961) Presidential address to the V international conference of soil mechanics and foundation engineering, Paris. *Proceedings 5° ICSMFE*. Dunod, Paris, Vol. III. pp. 49–50.

Squeglia, N., Stacul, N., Abed, A., Benz, T. & Leoni, M. (2018) m-PISE: A novel numerical procedure for pile installation and soil extraction. Application to the case of leaning tower of Pisa. *Computers and Geotechnics*, 102, 206–215.

Terracina, F. (1962) Foundations of the leaning tower of Pisa. *Geotechnique*, XII(4), 336–339.

Vasari, G. (1550) *Le vite de' più eccellenti architetti, pittori et scultori italiani*. Torrentino, Firenze, 2 Vol.

Vermeer, P.A., Neher, H.P., Vogler, U. & Bonnier, P.G. (2002) *3D Creep Analysis of the Leaning Tower of Pisa*. Unpublished research report, Stuttgart University.

Viggiani, C. & Pepe, M. (2005) Il sottosuolo della Torre. *Bollettino d'Arte, Vol. Spec. 2005: La Torre Restituita*, 2, 1–40.

Viggiani, C. & Squeglia, N. (2003) Electroosmosis to stabilize the leaning tower of Pisa. *Rivista Italiana di Geotecnica*, 37(1), 29–37.

Viggiani, C. & Squeglia, N. (2005a) Analisi della pendenza: origine ed evoluzione nel tempo. *Bollettino d'Arte, Vol. Spec. 2005: La Torre Restituita*, 1, 89–96.

Viggiani, C. & Squeglia, N. (2005b) Movimenti della fondazione della Torre. *Bollettino d'Arte, Vol. Spec. 2005: La Torre Restituita*, 2, 41–70.

Viggiani, C. & Squeglia, N. (2005c) Studio dei metodi per la stabilizzazione della Torre. *Bollettino d'Arte, Vol. Spec. 2005: La Torre Restituita*, 2, 139–154.

Viggiani, C. & Squeglia, N. (2005d) Studio dei metodi per la stabilizzazione della Torre. *Bollettino d'Arte, Vol. Spec. 2005: La Torre Restituita*, 2, 139–154.

Viggiani, C., Squeglia, N. & Pepe, M. (2005a) Analisi dei movimenti della Torre. *Bollettino d'Arte, Vol. Spec. 2005: La Torre Restituita*, 2, 71–94.

Viggiani, C., Squeglia, N. & Pepe, M. (2005b) Gli interventi di stabilizzazione temporanea. *Bollettino d'Arte, Vol. Spec. 2005: La Torre Restituita*, 2, 95–138.

Weckstein, L. (1999) *Through My Eyes: The 91st Infantry Division in the Italian Campaign 1942–1945*. Hellgate Press, ISBN:1-55571-497-8.

WeSI Geotecnica Srl. (2016) *3D Creep Analysis of the Leaning Tower of Pisa*. Unpublished report, Opera della Primaziale Pisana. p. 85.

Index

Note: Numbers in *italics* indicate a figure on the corresponding page.

T - #0276 - 071024 - C62 - 246/174/3 [5] - CB - 9780367469047 - Gloss Lamination